A2 Biology

Philip Allan Updates, an imprint of Hodder Education, an Hachette UK company, Market Place, Deddington, Oxfordshire OX15 0SE

Orders

Bookpoint Ltd, 130 Milton Park, Abingdon, Oxfordshire OX14 4SB
tel: 01235 827720   fax: 01235 400454   e-mail: uk.orders@bookpoint.co.uk

Lines are open 9.00 a.m.–5.00 p.m., Monday to Saturday, with a 24-hour message answering service. You can also order through our website: www.philipallan.co.uk

© Philip Allan Updates 2009
ISBN 978-0-340-99231-9

First published in 2006 as *Flashrevise Cards*

Impression number   5   4   3   2   1
Year   2014   2013   2012   2011   2010   2009

All rights reserved; no part of this publication may be reproduced, stored in a retrieval system, or transmitted, in any other form or by any means, electronic, mechanical, photocopying, recording or otherwise without either the prior written permission of Philip Allan Updates or a licence permitting restricted copying in the United Kingdom issued by the Copyright Licensing Agency Ltd, Saffron House, 6–10 Kirby Street, London EC1N 8TS.

Printed in Spain

Hachette UK's policy is to use papers that are natural, renewable and recyclable products and made from wood grown in sustainable forests. The logging and manufacturing processes are expected to conform to the environmental regulations of the country of origin.

P01537

Cellular respiration
1. Metabolism I
2. Metabolism II
3. Cellular respiration
4. Glycolysis
5. Aerobic respiration
6. Glycolysis and the link reaction
7. Krebs cycle
8. Electron transport chain
9. Mitochondria
10. Anaerobic respiration
11. The respiratory quotient
12. Measuring respiration

Photosynthesis
13. Photosynthesis
14. Light-dependent reaction I
15. Light-dependent reaction II
16. Light-independent reaction I
17. Light-independent reaction II
18. The role of chlorophyll
19. Chloroplasts
20. Leaf structure
21. Environmental factors
22. The rate of photosynthesis
23. Mineral nutrition

Ecology
24. Pyramids
25. Populations I
26. Populations II
27. Diversity
28. Succession
29. Agriculture and ecosystems I
30. Agriculture and ecosystems II
31. Chemical control of pests
32. Biological control of pests
33. Atmospheric pollution
34. Water pollution
35. Investigating ecosystems
36. Conservation: resources
37. Conservation: species

Genetics and evolution
38. Genes and alleles
39. Genotype and phenotype I
40. Genotype and phenotype II
41. Complex genetic relationships
42. Monohybrid inheritance
43. Sex determination
44. Dihybrid inheritance
45. Epistasis
46. Autosomal linkage
47. Sex linkage
48. Genetic crosses: techniques
49. Genetic crosses: analysing
50. Variation
51. Meiosis
52. Mutation

53 Polysomy and polyploidy
54 Population genetics
55 Evolution
56 Directional selection
57 Disruptive selection
58 Stabilising selection
59 Adaptations to the environment
60 Speciation: allopatric
61 Speciation: sympatric

Classification

62 Taxonomy
63 Animals and fungi
64 Plants, protoctista and prokaryotes

Homeostasis

65 Homeostasis
66 The regulation of blood glucose I
67 The regulation of blood glucose II
68 The regulation of blood glucose III
69 Thermoregulation
70 The liver: metabolism
71 The liver: excretion
72 The mammalian kidney I
73 The mammalian kidney II
74 The mammalian kidney III
75 The mammalian kidney IV

Coordination in plants and mammals

76 Plant growth substances I
77 Plant growth substances II
78 The phytochrome system
79 Non-sex hormones
80 Sex hormones
81 Nervous and chemical coordination

The nervous system

82 Neurones
83 Receptors and effectors
84 The nerve impulse I
85 The nerve impulse II
86 The synapse I
87 The synapse II
88 Muscles I
89 Muscles II
90 The eye: photoreceptors
91 The eye: visual pigments
92 The ear: balance
93 The ear: hearing
94 Reflexes
95 The autonomic nervous system
96 The central nervous system
97 Behaviour: innate
98 Behaviour: learned

Nutrition

99 Malnutrition
100 Overeating

A2 Biology
Cellular respiration

Metabolism 1

Q1 What is meant by the term metabolic pathway?

Q2 Distinguish between anabolism and catabolism.

Q3 List three advantages of metabolic pathways.

ANSWERS

1 ANSWERS

A1 A sequence of biochemical reactions controlled by enzymes

A2 Anabolic reactions are synthetic reactions that require energy. Catabolic reactions are breakdown reactions that release energy.

A3
- Metabolic reactions can proceed in a continuous manner. Equilibrium is never reached since the products become substrates of subsequent reactions.
- The stepwise nature of most catabolic pathways enables energy to be released in small, controlled amounts
- Each step is controlled by a specific enzyme and therefore each enzyme represents a point for control of the overall pathway

examiner's note All the chemical reactions that take place in cells make up metabolism. Substrates are converted into products during reactions catalysed by enzymes.

A2 Biology
Cellular respiration

Metabolism II

Q1 Explain what is meant by the following terms:
(a) phosphorylation (b) photolysis

Q2 Outline the action of each of the following enzymes, giving one example of their role in metabolism:
(a) hydrolases (b) oxidoreductases

Q3 During a metabolic reaction, molecule X gains electrons. Has molecule X been oxidised or reduced?

ANSWERS

2 ANSWERS

A1 (a) The addition of phosphate to a molecule
   (b) The splitting of a molecule using light energy

A2 (a) Hydrolases are enzymes that catalyse the hydrolysis of a molecule, i.e. the molecule is broken down by the addition of water. For example, the hydrolysis of glycogen releases glucose, to be used in respiration.
   (b) Oxidoreductases are enzymes that catalyse biochemical reactions involving oxidation and reduction. An example is cytochrome oxidase, which transfers electrons from cytochromes to oxygen (forming water) at the end of the electron transport chain in aerobic respiration.

A3 Reduced

***examiner's* note** In order to remember the details of oxidation and reduction, it is useful to recall the acronym OILRIG — Oxidation Is Loss, Reduction Is Gain (of electrons).

A2 Biology
Cellular respiration

Cellular respiration

Q1 Outline the structure of ATP (adenosine triphosphate).

Q2 Why is ATP important in metabolism?

Q3 List three roles of ATP in living organisms, other than its role in metabolic reactions.

3 ANSWERS

A1 ATP is a nucleotide, containing ribose as its pentose sugar and adenine as its nitrogenous base. Three phosphate groups are attached to the ribose subunit. The bond attaching the end phosphate group can be hydrolysed by ATPase to yield ADP (adenosine diphosphate) and a large quantity of energy.

A2 ATP supplies the energy needed for many metabolic reactions. It acts as an energy carrier in all living cells.

A3 ATP supplies the energy needed for:
- cell division
- active transport
- muscle contraction

examiner's **note** ATP has a number of roles in living organisms, including supplying the energy required for cell division, active transport and muscle contraction. It also plays an important role in metabolism, as it acts as the immediate supply of energy for anabolic processes, such as protein synthesis.

A2 Biology
Cellular respiration

4

Glycolysis

Q1 (a) Where in a cell does glycolysis take place?
(b) Is glycolysis a stage in aerobic respiration, anaerobic respiration, or both types of respiration?

Q2 (a) What is the main substrate used in glycolysis?
(b) List the three products of glycolysis.

Q3 What is the yield of ATP per molecule of glucose used in glycolysis?

ANSWERS

4 ANSWERS

A1 (a) In the cytoplasm of the cell
   (b) Both types of respiration

A2 (a) Glucose
   (b) Pyruvate, NAD (reduced) and ATP

A3 Two molecules of ATP

***examiner's* note** Glycolysis splits one molecule of glucose into two smaller molecules of pyruvate. The first stage of glycolysis is the phosphorylation of glucose to produce two molecules of triose phosphate. The triose phosphate is then oxidised to form two molecules of pyruvate. The net gain in terms of energy is two molecules of ATP.

A2 Biology
Cellular respiration

5

Aerobic respiration

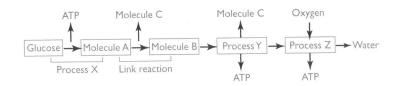

Q1 The flowchart summarises the process of aerobic respiration.
(a) Identify molecules A, B and C.
(b) Identify processes X, Y and Z.

Q2 Where in a cell does process Y take place?

5 ANSWERS

A1 (a) A = pyruvate
   B = acetylcoenzyme A
   C = carbon dioxide

(b) X = glycolysis
   Y = the Krebs cycle
   Z = the electron transport chain (oxidative phosphorylation)

A2 In the mitochondrial matrix

***examiner's* note** Cellular respiration is the process whereby cells break down glucose to release energy. This energy is used to produce ATP from ADP and inorganic phosphate. Aerobic respiration uses oxygen, but anaerobic respiration does not. Both types produce ATP, but anaerobic respiration only yields 5–6% of the ATP produced by aerobic respiration.

A2 Biology
Cellular respiration

Glycolysis and the link reaction

Q1 The link reaction can be summarised by the equation below. Identify molecules A, B and C.

molecule A + NAD (oxidised) → molecule B + molecule C + NAD (reduced)

Q2 NAD (nicotinamide adenine dinucleotide) acts as a hydrogen acceptor during respiration. Name the other hydrogen acceptor that is involved in aerobic respiration.

Q3 What happens to molecule B (see Q1) following the link reaction?

6 ANSWERS

A1 A = pyruvate
    B = acetylcoenzyme A
    C = carbon dioxide

A2 FAD (flavine adenine dinucleotide)

A3 Molecule B (acetylcoenzyme A) enters the Krebs cycle

***examiner's* note** Glycolysis (the first stage of aerobic respiration) produces three products:
- pyruvate — enters the matrix of the mitochondria for the link reaction and the Krebs cycle
- reduced NAD — used in the electron transport chain to generate further ATP
- ATP — used as an energy source

A2 Biology
Cellular respiration

Krebs cycle

Q1 The diagram shows some stages in the Krebs cycle. Identify molecules A and B.

Q2 State the number of carbon atoms present in each of compounds X, Y and Z.

Q3 Where in a cell does the Krebs cycle take place?

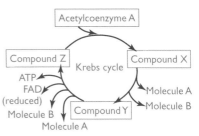

7 ANSWERS

A1 A = carbon dioxide
    B = NAD (reduced)

A2 X = 6 carbon atoms
    Y = 5 carbon atoms
    Z = 4 carbon atoms

A3 In the matrix of mitochondria

***examiner's* note** The Krebs cycle is the third stage of aerobic respiration. It happens once for each pyruvate molecule made in glycolysis, and it goes round twice for every glucose molecule that enters the respiration pathway. Products are either reused, released or used in the next stage of aerobic respiration:
- reused — oxaloacetate and coA
- released — carbon dioxide and ATP
- enter the electron transport chain — reduced NAD and reduced FAD

A2 Biology
Cellular respiration

Electron transport chain

Q1 For each of the four stages of aerobic respiration listed, state where in the cell it occurs: (a) glycolysis, (b) link reaction, (c) Krebs cycle and (d) electron transport chain.

Q2 Name the process by which the movement of electrons through the electron transport chain releases a large amount of energy to produce ATP.

Q3 Outline how the passage of electrons along the electron transport chain results in the production of ATP.

8 ANSWERS

A1 (a) The cell cytoplasm (b) The matrix of mitochondria
(c) The matrix of mitochondria (d) The cristae of mitochondria

A2 Oxidative phosphorylation

A3 The movement of electrons through the electron transport chain releases a large amount of energy, which is used to produce ATP by oxidative phosphorylation. The energy released by the transfer of electrons between carriers is used to pump protons (hydrogen ions) across the membrane. The return of the protons, through specialised protein pores in the membrane, is coupled to the synthesis of ATP.

examiner's note The purpose of the electron transport chain is to transfer the energy from molecules made in glycolysis, the link reaction and the Krebs cycle to produce ATP. The electron transport chain is where most of the ATP from aerobic respiration is produced.

A2 Biology
Cellular respiration

Mitochondria

Q1 Which figure is most likely to be the width of a mitochondrion?
10 nm   100 nm   1 µm   10 µm   100 µm

Q2 Suggest a type of cell that might contain large numbers of mitochondria.

Q3 Name the different stages of aerobic respiration that occur at these three sites: (a) cytoplasm, (b) mitochondrial matrix and (c) mitochondrial inner membrane (cristae).

ANSWERS

9 ANSWERS

A1 1 μm

A2 Muscle cells, metabolically active cells (e.g. liver cells or secretory cells) or cells that carry out active transport

A3 (a) Glycolysis
    (b) Link reaction and the Krebs cycle
    (c) Electron transport chain

***examiner's* note** Glycolysis takes place outside mitochondria, in the cytoplasm of the cell. If oxygen is present, pyruvate from glycolysis will enter the mitochondrial matrix for the link reaction and the Krebs cycle. Reduced coenzymes from these metabolic processes are oxidised on the inner membrane of the mitochondria to produce most of the ATP made during aerobic respiration.

A2 Biology
Cellular respiration

Anaerobic respiration

The diagram summarises the process of anaerobic respiration.

Q1 Identify compound A.

Q2 What groups of living organisms are represented by B and C?

Q3 Why is it important that pyruvate acts as a hydrogen acceptor in anaerobic respiration?

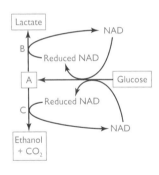

10 ANSWERS

A1 A = pyruvate

A2 B = animals; C = plants and microorganisms

A3 So that reduced NAD can be re-oxidised to oxidised NAD. This allows further breakdown of glucose to pyruvate, and so glycolysis continues and ATP is generated.

***examiner's* note** There are four key differences between aerobic and anaerobic respiration:
- aerobic respiration needs oxygen, anaerobic respiration does not
- aerobic respiration takes place in the cytoplasm and mitochondria, anaerobic respiration takes place in the cytoplasm only
- aerobic respiration involves one metabolic pathway, but anaerobic respiration can involve either of two metabolic pathways — alcoholic fermentation or lactate fermentation
- aerobic respiration produces 38 ATP for every glucose, but anaerobic respiration produces only 2 ATP for every glucose

A2 **Biology**
Cellular respiration

The respiratory quotient

Q1 What is meant by the term respiratory quotient (RQ)?

Q2 Match each respiratory substrate to its corresponding RQ value.

Lipid	1.0
Protein	0.7
Carbohydrate	0.9

Q3 What does an RQ value of 0.85 suggest about the type of respiratory substrate being used?

11 ANSWERS

A1 The volume of carbon dioxide produced during respiration divided by the volume of oxygen used in a set period of time

A2 Lipid — 0.7
Protein — 0.9
Carbohydrate — 1.0

A3 An RQ value of 0.85 indicates that a mixture of carbohydrates and lipids is being used in respiration. Protein is rarely used as an energy source except in cases of food deprivation.

***examiner's* note** High RQs (greater than 1.0) often mean that an organism is short of oxygen and is having to respire anaerobically as well as aerobically. Plants sometimes have a low RQ because some of the carbon dioxide released in respiration is used for photosynthesis.

A2 Biology
Cellular respiration

12

Measuring respiration

Q1 What is measured by a respirometer?

Q2 What is the role of the sodium hydroxide solution in a respirometer?

Q3 Why are the experimental and control tubes in a respirometer kept in a water bath at constant temperature?

ANSWERS

12 ANSWERS

- **A1** The rate of respiration, as indicated by the rate of oxygen consumption
- **A2** To absorb carbon dioxide so that any change in the volume of air is due to the consumption of oxygen
- **A3** To ensure that any changes in the volume of gas in the tubes is due to respiration and not to changes in temperature

***examiner's* note** Temperature affects the rate of respiration. The rate doubles for every rise of 10°C up to around 45°C, when it slows due to the denaturation of enzymes. The effect of temperature on the rate of respiration can be measured using a respirometer — the apparatus is placed in a thermostatically-controlled water bath and the temperature of the water is varied.

A2 Biology
Photosynthesis

Photosynthesis

The diagram summarises the chemical reactions that take place during photosynthesis.

Q1 Define the term photosynthesis.

Q2 Identify molecules A, B, C and D.

Q3 Identify processes X and Y.

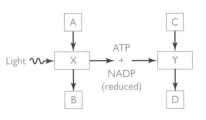

13 ANSWERS

A1 The process by which organic compounds are synthesised from carbon dioxide and water using light energy

A2 A = water
B = oxygen
C = carbon dioxide
D = carbohydrate (glucose)

A3 X = light-dependent reaction
Y = light-independent reaction

***examiner's* note** The overall equation for photosynthesis is as follows:

$$6CO_2 + 6H_2O + \text{light energy} \rightarrow C_6H_{12}O_6 + 6O_2$$

The process can be split into two stages: the *light-dependent* reaction and the *light-independent* reaction.

A2 Biology
Photosynthesis

14

Light-dependent reaction I

Q1 (a) Outline what is meant by the light-dependent reaction.
(b) Where does the light-dependent reaction take place?

Q2 What are the two processes involved in the light-dependent reaction?

Q3 In the diagram showing cyclic photophosphorylation, identify X and Y.

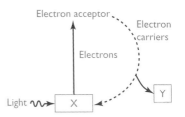

14 ANSWERS

A1 (a) The process by which light energy is absorbed by chlorophyll in the photosystems and used to produce ATP and reduced NADP during photosynthesis
(b) On the grana of chloroplasts

A2 • Cyclic photophosphorylation
• Non-cyclic photophosphorylation

A3 X = photosystem 1
Y = ATP

***examiner's* note** Cyclic photophosphorylation uses only photosystem 1. Energy from high-energy electrons is used to generate ATP, which is used in the light-independent reaction.

A2 Biology
Photosynthesis

Light-dependent reaction II

Q1 List the four substrates of the light-dependent reaction.

Q2 List the three products of the light-dependent reaction.

Q3 Distinguish between cyclic and non-cyclic photophosphorylation.

15 ANSWERS

A1
- Inorganic phosphate
- NADP (oxidised)
- Water
- ADP

A2
- ATP
- Oxygen
- NADP (reduced)

A3 In both cases, light energy causes the emission of a high-energy electron from chlorophyll and ATP is generated. In cyclic photophosphorylation, the electron is passed back to the chlorophyll molecule. In non-cyclic photophosphorylation, the electron is used to reduce NADP. Chlorophyll has its electron replaced by the photolysis of water, resulting in the production of oxygen.

***examiner's* note** Non-cyclic photophosphorylation uses both photosystem 1 and photosystem 2. Energy from high-energy electrons is used to generate ATP and to reduce NADP, which are used in the light-independent reaction.

A2 Biology
Photosynthesis

16

Light-independent reaction I

Q1 Where does the light-independent reaction take place?

Q2 State whether the following substances are substrates or products of the light-independent reaction.
   (a) ATP
   (b) NADP (oxidised)
   (c) Carbohydrate
   (d) Ribulose bisphosphate
   (e) Carbon dioxide

Q3 List the steps involved in the reduction of carbon dioxide to starch in the light-independent reaction.

ANSWERS

16 ANSWERS

A1 In the stroma of chloroplasts

A2 (a) Substrate   (c) Product   (e) Substrate
    (b) Product   (d) Substrate

A3
- Carbon dioxide combines with ribulose bisphosphate to form glycerate 3-phosphate
- Glycerate 3-phosphate is reduced to glyceraldehyde 3-phosphate
- Glyceraldehyde 3-phosphate is converted to glucose
- Glucose subunits are joined together by condensation reactions to form starch

***examiner's* note** To ensure that there are always sufficient supplies of ribulose bisphosphate to combine with carbon dioxide, five out of every six molecules of triose phosphate produced in the Calvin cycle (part of the light-independent cycle) are used to regenerate ribulose bisphosphate instead of being used to make hexose sugars, thereby keeping the cycle going.

A2 Biology
Photosynthesis

Light-independent reaction II

The diagram shows the Calvin cycle.

Q1 How many carbon atoms has:
(a) ribulose bisphosphate?
(b) glycerate 3-phosphate?

Q2 Name molecule X.

Q3 Explain what would happen if all the glyceraldehyde 3-phosphate was converted to other carbohydrates.

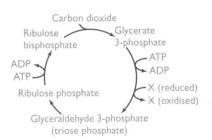

17 ANSWERS

A1 (a) Five
(b) Three

A2 NADP

A3 The process would stop, since the acceptor molecule for carbon dioxide, ribulose bisphosphate, must be regenerated in order to fix the carbon dioxide from the atmosphere

***examiner's* note** The light-independent reaction is dependent on ATP and reduced NADP produced in the light-dependent reaction. Therefore, in the absence of light, the light-independent reaction will eventually stop.

A2 Biology
Photosynthesis

The role of chlorophyll

Q1 What is an absorption spectrum?

Q2 Why are most leaves green in colour?

Q3 What is an action spectrum?

Q4 The action spectrum for photosynthesis is similar to the absorption spectrum for chlorophyll a. What does this suggest?

ANSWERS

ANSWERS

A1 A graph showing the relative amounts of light of different wavelengths that are absorbed by a pigment

A2 Most leaves appear green because chlorophyll absorbs light strongly at the blue and red ends of the spectrum, and reflects green light

A3 A graph showing the relative amounts of light of different wavelengths that are used in a particular process, such as photosynthesis

A4 It suggests that most of the light absorbed by chlorophyll is used in photosynthesis

***examiner's* note** Photosystems are made up of chlorophyll a, accessory pigments (e.g. chlorophyll b and carotenoids) and proteins. The proteins hold the pigment molecules in the best position for absorbing light energy and for transferring this energy to the reaction centre of the photosystem — for use in the light-dependent reaction.

A2 Biology
Photosynthesis

19

Chloroplasts

Q1 List six structures found in chloroplasts.

Q2 Which figure is most likely to be the width of a chloroplast?
25 nm   250 nm   2.5 µm   25 µm   250 µm

Q3 State where in a chloroplast the following stages of photosynthesis take place:
(a) light-dependent reaction
(b) light-independent reaction

ANSWERS

19 ANSWERS

A1 Any six of:
- Ribosomes
- Stroma
- Grana
- Envelope (outer and inner membrane)
- DNA
- Lipid droplets
- Starch grains
- Thylakoid

A2 2.5 µm

A3 (a) Grana
(b) Stroma

***examiner's* note** Chloroplasts are well adapted to carry out the function of photosynthesis. The grana have a large surface area for the light-dependent reaction to take place. Photosystems 1 and 2 are very efficient at capturing light energy. The stroma contains enzymes for the Calvin cycle and stores carbohydrate as starch grains.

A2 Biology
Photosynthesis

Leaf structure

Q1 In which type of leaf cell does the majority of photosynthesis occur?

Q2 What is the name given to the small pores found predominantly on the underside of dicotyledonous leaves?

Q3 Describe how a leaf is adapted for efficient photosynthesis.

ANSWERS

ANSWERS

A1 Palisade mesophyll cells

A2 Stomata

A3
- The waxy cuticle prevents dehydration but is transparent, thereby allowing the passage of light
- Leaves are usually broad, thin and flat, thereby maximising light absorption and permitting the carbon dioxide which enters the leaf through the stomata to reach the palisade cells easily
- Palisade cells are packed together, close to the top of the leaf
- Veins in the leaf contain xylem vessels which bring water from the roots, and phloem vessels which transport sugars made in photosynthesis to the rest of the plant

***examiner's* note** The stomata and air spaces in the spongy mesophyll of the leaf permit efficient gas exchange between palisade cells and the environment — carbon dioxide enters the cells and oxygen is removed. Palisade cells contain many chloroplasts for efficient photosynthesis.

A2 Biology
Photosynthesis

21

Environmental factors

Q1 Explain what is meant by Blackman's law of limiting factors.

Q2 List the three main factors affecting the rate of photosynthesis.

Q3 Under normal environmental conditions, which of these is a limiting factor at very low rates of photosynthesis?

ANSWERS

A1 When a process is controlled by a number of factors, the factor in least supply will limit the rate of the process

A2
- Light intensity
- Carbon dioxide concentration
- Temperature

A3 Light intensity

***examiner's* note** When light intensity is no longer limiting the rate of photosynthesis, increasing temperature and/or the concentration of carbon dioxide may cause an increase in the rate of photosynthesis. The point on the graph where the rate of photosynthesis is the same as the rate of respiration is known as the 'compensation point'. At this point, the amount of carbon dioxide produced by respiration is exactly equal to the amount of carbon dioxide being used in photosynthesis.

A2 Biology
Photosynthesis

22

The rate of photosynthesis

Q1 Name the aquatic plant that is commonly used to measure the rate of photosynthesis.

Q2 List three factors, other than light intensity, that affect the rate of photosynthesis.

Q3 Outline how you would measure the effect of light intensity on the rate of photosynthesis.

ANSWERS

A1 Canadian pondweed (*Elodea*)

A2 Any three of:
- Light wavelength
- Carbon dioxide concentration
- Temperature
- Water availability

A3 A source of light is placed at different distances from the plant and the volume of oxygen produced is measured. It is important to vary only light intensity and keep all other factors constant (e.g. light wavelength, temperature) when measuring the effect on oxygen production.

***examiner's* note** In the UK, the optimum conditions for most plants are:
- high light intensity
- temperature of around 25°C
- a constant supply of water
- carbon dioxide at a concentration of about 0.4%

A2 Biology
Photosynthesis

23

Mineral nutrition

Q1 In addition to carbon dioxide and water, which two essential elements are required by flowering plants to produce nucleic acids?

Q2 Why do plants need to take up magnesium ions from the soil?

Q3 Why can waterlogged soil have an adverse effect on the health of plants?

ANSWERS

ANSWERS

A1 Nitrogen and phosphorous

A2 To use in the synthesis of chlorophyll and as an activator for many enzymes

A3 Waterlogged soil can reduce the uptake of mineral ions, as this type of soil is poorly aerated. The reduced supply of oxygen to roots inhibits the efficient active transport of mineral ions into plants.

***examiner's* note** Mineral ions are taken up by root cells by diffusion and active transport, and moved through the plant as solutes in water. Active transport is required since diffusion is not a selective process — plants often only need certain ions and not others. Active transport allows plants to absorb the necessary mineral ions against the concentration gradient, as well as to pump out those that are not required.

A2 Biology
Ecology

Pyramids

Q1 What is a pyramid of biomass?

Q2 The organisms in an oak wood are shown in the table. Draw a pyramid of numbers for the oak wood.

Number	Organisms
200	Oak trees
100 000	Primary consumers
800 000	Secondary consumers
5000	Tertiary consumers

Q3 Draw a pyramid of energy for the oak wood.

ANSWERS

A1 A pyramid of biomass shows the mass of organisms at each trophic level in a food chain

A2

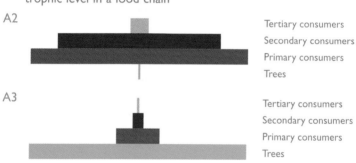

Tertiary consumers
Secondary consumers
Primary consumers
Trees

A3

Tertiary consumers
Secondary consumers
Primary consumers
Trees

***examiner's* note** Not all pyramids of numbers have the characteristic upright pyramid shape. The shape varies according to the *size* of the organisms involved. Pyramids of biomass and energy are more likely to have a simple pyramid shape.

A2 Biology
Ecology

Populations 1

Q1 What is meant by the term population?

Q2 Distinguish between population size and population density.

Q3 Distinguish between density-dependent and density-independent factors.

ANSWERS

25 ANSWERS

A1 A population is a group of individuals of the same species occupying a certain area

A2 Population size refers to the total number of individuals present in a given area, whereas population density is a measure of the number of individual organisms per unit area

A3 Density-dependent factors are factors limiting the size of a population whose effects are proportional to the density of the population. These factors tend to be biotic, for example, food supply (for both predators and prey) and infectious disease. Density-independent factors are factors limiting the size of a population whose effects are independent of the density of the population. These factors tend to be abiotic, for example, temperature and water availability.

***examiner's* note** Population growth is limited by environmental factors. Many limiting factors are abiotic, such as temperature, pH, oxygen availability and the supply of mineral nutrients.

A2 Biology
Ecology

Populations II

Q1 List the conditions under which you would expect a population to grow at its maximum rate.

Q2 Explain what is meant by the following terms:
(a) carrying capacity
(b) environmental resistance

Q3 Distinguish between interspecific competition and intraspecific competition.

ANSWERS

A1
- A plentiful supply of food and water
- A good supply of suitable habitats
- An absence of infectious diseases
- An absence of predators
- A suitable pH
- Good light intensity
- A suitable temperature

A2 (a) The maximum population size that can be sustained by a particular habitat

(b) A combination of density-dependent and density-independent factors that prevent populations from growing beyond the carrying capacity

A3 Interspecific competition is competition for resources between members of different species, whereas intraspecific competition is competition for resources between members of the same species

***examiner's* note** There are two important interactions that limit population size — competition and predation. Population sizes of predators and prey are interlinked.

A2 Biology
Ecology

Diversity

Q1 What is meant by the term diversity?

Q2 Outline the distinguishing features of an ecosystem with:
(a) low diversity
(b) high diversity

Q3 Calculate the diversity index of ecosystem X, given that $N = 18$ and $\Sigma n(n-1) = 42$.

ANSWERS

27 ANSWERS

A1 A measure of the number of different species present in an ecosystem

A2 (a) Low diversity is common in extreme ecosystems such as deserts, tundra and salt marshes. In these areas, plant and animal populations are mainly affected by abiotic factors and the diversity index is low.
(b) High diversity is common in ecosystems that are usually mature, natural (i.e. not created by human activity) and have environmental conditions that are not too hostile. In these ecosystems, populations are mostly affected by biotic factors and the diversity index is high.

A3 diversity index = $[N(N-1)/ \Sigma n(n-1)]$; so diversity index = $[18(17)/42] = 7.3$

examiner's note Three key factors are used to define the diversity of an ecosystem:
- species frequency — how abundant a species is in an area
- species richness — the total number of different species in an area
- percentage cover (plants only) — how much of the surface is covered by a particular plant species

A2 Biology
Ecology

Succession

Q1 What is succession?

Q2 Distinguish between primary succession and secondary succession.

Q3 Place the following stages of succession in the correct order, from earliest to latest (climax community):

grass community, mature woodland, marram grass, sand dunes, shrub community

ANSWERS

28 ANSWERS

A1 The process by which plant communities gradually develop on bare land. The end point of succession is the development of a stable climax community.

A2 Primary succession occurs on land where there is no soil and no living organisms, such as that created by a volcanic eruption. Secondary succession occurs when most of the living organisms in an area have been destroyed but the soil and some living organisms remain. An example is woodland that has been burned by a forest fire.

A3 (i) Sand dunes (earliest)
(ii) Marram grass
(iii) Grass community
(iv) Shrub community
(v) Mature woodland (climax community)

***examiner's* note** Diversity increases and species change as succession progresses.

A2 Biology
Ecology

Agriculture and ecosystems I

Q1 State which type of farming (extensive or intensive) is characterised by each of the following features:
(a) labour intensive
(b) little use of pesticides
(c) little capital investment required
(d) production of a single crop (monoculture) or animal species

Q2 Give an advantage and a disadvantage of intensive farming methods.

Q3 What problem is caused when fertilisers enter rivers and lakes?

ANSWERS

29 ANSWERS

A1 (a) Extensive
(b) Extensive
(c) Extensive
(d) Intensive

A2 Advantage — boosts food production

Disadvantage — (any one of) destruction of habitats, pollution, a reduction in diversity

A3 Eutrophication

***examiner's* note** Traditional farming methods are sometimes referred to as *extensive* food production. Modern agriculture is characterised by *intensive* farming.

A2 Biology
Ecology

Agriculture and ecosystems II

Q1 What is intercropping?

Q2 Explain how planting the following helps to minimise the negative effects of agriculture on ecosystems:
(a) hedgerows
(b) trees
(c) legumes

Q3 Instead of chemical pesticides, what alternative method can farmers use to get rid of pests?

ANSWERS

A1 The practice of growing two or more crops in the same field at the same time. This uses space efficiently and makes maximum use of soil nutrients. Intercropping also encourages biodiversity.

A2 (a) Hedgerows increase the number of habitats available for colonisation and thus species diversity
(b) Trees act as wind shields and also prevent soil erosion
(c) Legumes (e.g. peas, beans and clover) naturally restore nitrates to the soil

A3 Biological control

examiner's note Farming can be intensive or extensive. Intensive farming methods can adversely affect the environment, but farmers can use a range of strategies to minimise the negative effects of agriculture.

A2 **Biology**
Ecology

Chemical control of pests

Q1 List three different categories of pesticide.

Q2 Give two advantages of chemical control.

Q3 Give three disadvantages of chemical control.

ANSWERS

31 ANSWERS

A1
- Herbicides
- Insecticides
- Fungicides

A2
- It is often very effective at completely eradicating the pest species
- It is relatively fast compared with biological control

A3 Any three of:
- It may be toxic to non-target organisms
- It may be non-biodegradeable and persist in the environment
- It may bioaccumulate in living organisms (becoming particularly concentrated in animals found at the top of food chains)
- Its frequent use can result in the target pest population becoming resistant to its effects

***examiner's* note** Pesticides can be classified as either *contact* or *systemic*. Contact pesticides kill organisms when they come into direct contact with them. Systemic pesticides have to be absorbed into the organism in order to kill it.

A2 Biology
Ecology

32

Biological control of pests

Q1 What is meant by the term biological control?

Q2 Give three advantages of biological control.

Q3 Give three disadvantages of biological control.

ANSWERS

ANSWERS

A1 The use of a parasite or predator to control the number of pest organisms in a particular area (e.g. aphids are controlled by the use of ladybirds). Biological control also includes breeding pest-resistant crops.

A2
- It is very specific
- Pests do not usually become resistant to biological control
- It does not pollute the environment

A3
- It is relatively slow compared with chemical control
- Pests are never completely eliminated, so there is always some damage to crops
- The control organism may become a pest in its own right

***examiner's* note** In recent years, farmers have started to use integrated pest management, which combines chemical and biological methods.

A2 Biology
Ecology

Atmospheric pollution

Q1 What is pollution?

Q2 Name two gases that contribute to the formation of acid rain.

Q3 (a) Name two gases that contribute to the greenhouse effect.
(b) Explain how the greenhouse effect may lead to global warming.
(c) Suggest one environmental consequence of global warming.

33 ANSWERS

A1 Pollution may be defined as a change in the abiotic or biotic characteristics of the environment as a result of human activities which introduce harmful substances into the atmosphere and water supplies

A2 Sulphur dioxide and nitrogen oxides

A3 (a) Carbon dioxide and methane
(b) The 'greenhouse gases' prevent some of the sun's radiation heat from leaving the atmosphere. As the concentration of these gases increases, more heat is trapped in the atmosphere, leading to an increase in the average temperature of the earth — global warming.
(c) Melting of the polar icecaps leading to rising sea levels and flooding

***examiner's* note** The main atmospheric pollutants are sulphur dioxide, nitrogen oxides, carbon dioxide and methane. These pollutants lead to acid rain and global warming. Chlorofluorocarbons (CFCs) are other atmospheric pollutants that damage the ozone layer.

A2 Biology
Ecology

Water pollution

Q1 What is eutrophication?

Q2 Suggest two possible causes of eutrophication.

Q3 Outline how eutrophication leads to reduced species diversity in a river or lake.

ANSWERS

A1 Eutrophication is the decrease in biodiversity resulting from the pollution of a river or lake. The pollutants stimulate the growth of algae (algal bloom) in the water, which eventually die leading to an increase in the respiration of aerobic decomposers (e.g. bacteria).

A2 Fertilisers and sewage entering rivers or lakes

A3 The algal bloom caused by eutrophication and the subsequent increase in aerobic respiration by the decomposing bacteria results in an increased biochemical oxygen demand and subsequent reduction in the oxygen concentration in the water. Consequently, many aquatic organisms, such as fish, die due to lack of oxygen, reducing species diversity in the river or lake.

***examiner's* note** Indicator species can be used to check water quality. Some organisms can only survive in clean water, so their presence indicates a lack of pollution. The presence of other organisms indicates water polluted with organic matter.

A2 Biology
Ecology

Investigating ecosystems

Q1 Describe how the following tools are used to investigate ecosystems:
(a) quadrats
(b) transects

Q2 Outline the methods used to trap animals.

Q3 Calculate the population size of snails in a habitat given the following information from a mark-release-recapture study: $n_1 = 24$, $n_2 = 27$ and $n_m = 8$.

35 ANSWERS

A1 (a) Quadrats are used to sample areas of plant cover and can be point or area quadrats of different sizes. They provide information about percentage cover, species richness and diversity.
(b) Transects are used to survey an area and are useful in investigating trends, such as the distribution of organisms on a rocky shore. They may be along a single line or in a belt of land.

A2 Methods include: traps to catch small mammals; nets to catch flying insects and aquatic animals; pitfall traps and pooters to catch certain insects; Tullgren funnels to extract small animals from soil samples

A3 Population size = $(n_1 \times n_2)/n_m = (24 \times 27)/8 = 81$

***examiner's* note** The accuracy of the mark-release-recapture technique depends on three main assumptions:
- the marked sample has mixed evenly back into the total population
- marking has not affected the individuals' chances of survival (e.g. more visible to predators)
- changes in population size due to births, deaths and migration are small

A2 Biology
Ecology

Conservation: resources

Q1 What is conservation?

Q2 List the key resources required for human life.

Q3 Water treatments are used to recycle water. Put these stages in the correct order, starting with sewage and ending with drinking water:

sewage, aeration, filtration, chemical treatment, sedimentation and fermentation, drinking water

ANSWERS

ANSWERS

A1 The protection and maintenance of natural resources

A2
- Food and water
- Other species
- Energy
- Land
- Minerals and other raw materials

A3
(i) Sewage
(ii) Filtration
(iii) Sedimentation and fermentation
(iv) Aeration
(v) Chemical treatment
(vi) Drinking water

examiner's **note** Conservation is not the same as preservation. Natural systems are dynamic, with continual cycles and changes, so conservation needs active management. It may also involve repair and reclamation after earlier environmental damage.

A2 Biology
Ecology

Conservation: species

Q1 What are the two main factors that contribute to the destruction of natural habitats?

Q2 Identify two measures that have been put in place to promote habitat conservation in the UK.

Q3 Name two environments in which species can be conserved outside their natural habitats.

ANSWERS

A1
- Expansion of human populations
- Over-exploitation of resources, e.g. deforestation

A2 Any two of:
- National parks
- Sites of Special Scientific Interest (SSSIs)
- Environmentally Sensitive Areas (ESAs)
- Wildlife reserves

A3 Zoos and botanical gardens

examiner's note Tropical rainforests are important for the whole biosphere due to their rich species diversity and stabilising effect on the global climate. Conservation measures for rainforests include careful management, developing alternatives to use in place of tropical hardwoods and laying aside conservation areas that are completely free from logging.

A2 Biology
Genetics and evolution

Genes and alleles

Q1 What is a gene?

Q2 What is the difference between genes and alleles?

Q3 Distinguish between the terms dominant and recessive.

ANSWERS

38 ANSWERS

A1 A gene is a section of DNA that codes for the production of a particular polypeptide or protein. A gene may also be defined as a unit of heredity that is found at a specific position on a chromosome.

A2 Different forms of a particular gene are known as alleles. A gene controls a character, such as eye colour, and alleles control different forms of this character, such as blue, green or brown eyes.

A3 Dominant refers to any allele that is always expressed in the phenotype of an organism.

Recessive refers to any allele that is only expressed in the phenotype of an organism if the dominant allele is not present.

***examiner's* note** Genes are units of heredity that control the production of proteins. Alleles are the different forms of a particular gene. Each gene can have one, two or more alleles, and these alleles may be dominant, recessive or codominant.

A2 Biology
Genetics and evolution

Genotype and phenotype I

Q1 Distinguish between the terms homozygous and heterozygous.

Q2 For a gene given the symbol R, name the alleles present in:
(a) a heterozygous organism
(b) a homozygous recessive organism

Q3 Insert the words **genotype**, **phenotype** and **environment** into the word equation to show the relationship between these terms.

.... + ... = ...

A1 Homozygous refers to a genotype in which the alleles are the same, e.g. TT and tt are homozygous genotypes.

Heterozygous refers to a genotype in which the alleles are different, e.g. Tt is a heterozygous genotype.

A2 (a) Rr
(b) rr

A3 genotype + environment = phenotype

***examiner's* note** Much variation in characteristics (phenotype) is due to differences in genotype, i.e. the alleles present. However, the environment can also have a significant effect on variation.

A2 Biology
Genetics and evolution

Genotype and phenotype II

Q1 Identify the terms referred to in the following definitions:
(a) Any allele that is always expressed in the phenotype of an organism.
(b) The physical expression of the genotype of an organism.

Q2 (a) Describe the relationship between genotypic and phenotypic variation.
(b) What can cause variation between genetically identical individuals?

Q3 Explain how studies of monozygotic twins show the effects of genes and the environment.

ANSWERS

40 ANSWERS

A1 (a) Dominant
(b) Phenotype

A2 (a) Variation in phenotype is a combination of variation in genotype plus variation in the environment
(b) Environmental factors

A3 Monozygotic twins have identical alleles because they develop from the same fertilised egg. This means that if there are any differences in their characteristics, they must be due to the environment.

***examiner's* note** Phenotypic variation is due to a combination of genotypic variation and variation in the environment. Some characteristics, such as ABO blood groups, are independent of environmental influences. Others, such as whether someone has their ears pierced, are totally environmental. Most characteristics are a combination of genes (nature) and environment (nurture).

A2 Biology
Genetics and evolution

41

Complex genetic relationships

Q1 What is the term used to describe any two alleles which are both expressed in the phenotype of a heterozygous organism?

Q2 (a) How many genes are responsible for ABO blood groups?
(b) How many alleles are responsible for ABO blood groups?

Q3 Complete the table to show the possible genotypes and phenotypes associated with ABO blood groups.

Genotype	Phenotype (blood group)
$I^A I^A$ or $I^A I^O$	
	O
	AB
	B

ANSWERS

41 ANSWERS

A1 Codominant

A2 (a) One (b) Three

A3

Genotype	Phenotype (blood group)
$I^A I^A$ or $I^A I^O$	A
$I^O I^O$	O
$I^A I^B$	AB
$I^B I^B$ or $I^B I^O$	B

***examiner's* note** One example of codominance in humans involves the allele for sickle-cell anaemia. Usually, people have two alleles for normal haemoglobin ($H^N H^N$), but those who suffer from sickle-cell anaemia have two alleles for the disease ($H^S H^S$). Their red blood cells are sickle-shaped and do not carry sufficient oxygen. Heterozygous people ($H^N H^S$) have an in-between phenotype, called the sickle-cell trait. Some red blood cells are normal and some are sickle-shaped. The two alleles are codominant because they are both expressed in the phenotype.

A2 Biology
Genetics and evolution

Monohybrid inheritance

Q1 What is meant by the term monohybrid inheritance?

Q2 Flower colour in snapdragons is determined by a gene with two codominant alleles: C^R and C^W. Flowers with the genotype $C^R C^R$ are red, those with $C^W C^W$ are white and those with $C^R C^W$ are pink. What would be the result of interbreeding two pink flowered snapdragons?

Q3 Which blood group genotypes would result in the offspring from parents with genotypes $I^A I^O$ and $I^B I^O$, respectively?

ANSWERS

- **A1** The inheritance of a single characteristic. This is determined by a single gene which can have one or more different alleles. Two alleles will determine the type of characteristic inherited.
- **A2** C^WC^W (white); C^WC^R (pink); and C^RC^R (red) in the ratio of 1 white : 2 pink : 1 red
- **A3** Each offspring would have an equal probability (25%) of having a genotype I^AI^B (blood group AB), I^AI^O (blood group A), I^BI^O (blood group B) or I^OI^O (blood group O)

***examiner's* note** A test cross can be used to determine if the genotype of an organism showing the dominant phenotype is homozygous or heterozygous. For example, if a pea plant that produces yellow seeds (BB or Bb) is subjected to a test cross with a plant producing green seeds (bb), the results will tell us the unknown genotype. If the resulting plants produce both yellow and green seeds in approximately equal numbers, the parent plant producing yellow seeds must have been Bb (heterozygous). If, however, no green seeds are produced, the parent plant must have been BB (homozygous).

A2 Biology
Genetics and evolution

43

Sex determination

Q1 Why are female mammals known as the homogametic sex?

Q2 Why are there two types of sperm cells?

Q3 Explain the following statements:
(a) In humans, it is the man that determines the sex of the baby.
(b) In any one year, equal numbers of male and female births would be expected.

43 ANSWERS

A1 All eggs (gametes) produced by a female contain an X chromosome

A2 Male mammals have one X and one Y chromosome (XY) and sperm (gametes) may contain either an X chromosome or a Y chromosome

A3 (a) Women produce eggs that can only contain an X chromosome. Men produce sperm containing either an X chromosome or a Y chromosome. Therefore, the sex of the baby depends on which type of sperm fertilises the egg — an X sperm will produce a female (XX) and a Y sperm will produce a male (XY).
(b) Men produce approximately equal numbers of X and Y chromosome-containing sperm. Therefore, we would expect approximately equal numbers of male and female births.

***examiner's* note** The Y chromosome is smaller than the X chromosome and carries fewer genes. So most genes carried on the sex chromosomes are carried only on the X chromosome. These genes are sex-linked.

A2 Biology
Genetics and evolution

44

Dihybrid inheritance

Q1 What is meant by the term dihybrid inheritance?

Q2 Coat colour and fur length in rats are controlled by two unlinked genes. The alleles determining coat colour are B (black, dominant) and b (brown, recessive). The alleles determining fur length are L (long, dominant) and l (short, recessive). What combination of alleles would produce a rat with long black fur?

Q3 What would be the offspring from a cross between a rat of genotype bbLl and a rat with short brown fur?

ANSWERS

ANSWERS

A1 The inheritance of two different characteristics. These will be determined by two genes, each of which can have one or more different alleles. Two alleles for each gene will determine the type of characteristic inherited.

A2 BB or Bb with LL or Ll

A3 Offspring of genotypes bbLl and bbll in approximately equal numbers. In other words, rats with long brown fur and rats with short brown fur in a ratio of 1:1.

***examiner's* note** Dihybrid inheritance shows how two different genes are inherited. Each gene gives a 3:1 ratio in the F_2 generation, but because the two genes do this independently, it makes a 9:3:3:1 ratio overall.

A2 Biology
Genetics and evolution

Epistasis

Q1 What is meant by the term epistasis?

Q2 Daffodils may have yellow or cream flowers. Yellow results from the dominant allele Y and cream from the recessive allele y. However, flower colour is also affected by a second pair of alleles. The dominant allele C allows the yellow or cream colour to develop, but in the homozygous recessive (cc), no colour will develop and the flower will be white. What would be the result if two plants of genotype YyCc were crossed?

ANSWERS

45 ANSWERS

A1 Epistasis is a type of gene interaction in which one gene controls the expression of another gene

A2 Offspring in the ratio of:
9 yellow flowers : 3 cream flowers : 4 white flowers

***examiner's* note** Sometimes two different genes control the same characteristic and they interact in the phenotype. This is called polygenic inheritance. One example of this is when one gene can prevent the other one from being expressed. This is known as epistasis.

A2 Biology
Genetics and evolution

Autosomal linkage

Q1 What is meant by the term autosomal linkage?

Q2 Why does an unusual genetic ratio suggest autosomal linkage?

Q3 Explain the results of a cross between two heterozygous tall pea plants with purple flowers which gave the following ratio of phenotypes:

| 152: | 38: | 41: | 24 |
| tall purple | tall white | short purple | short white |

ANSWERS

A1 Linkage describes the relationship between different genes located on the same chromosome. Genes on a particular chromosome will normally be inherited together, as a linkage group, unless they are separated during crossing-over.

A2 The genes are linked, but are separated occasionally by crossing-over during meiosis. As a result, no simple genetic ratio (e.g. 3:1 or 9:7) would be expected.

A3 The ratio is approximately 25:6:7:4. This is significantly different to the 9:3:3:1 ratio expected. It occurs because the genes for height and colour are linked.

***examiner's* note** Autosomal linkage causes some genes to be inherited together, so that dihybrid inheritance ratios appear to become monohybrid ratios, e.g. 3:1 instead of 9:3:3:1. The issue is complicated by crossing-over between genes during meiosis — so that total linkage is rare and some unusual phenotypic ratios can arise in offspring.

A2 Biology
Genetics and evolution

47

Sex linkage

Q1 Distinguish between the terms sex linkage and autosomal linkage.

Q2 In cats, the gene for coat colour is carried on the X chromosome. There are two codominant alleles, B (black) and G (ginger). A cat with both B and G alleles is a colour known as tortoiseshell. What would be the expected result of a cross between a tortoiseshell cat and a ginger cat?

Q3 What would be the result of a cross between a normal human male and a human female who was a carrier for haemophilia?

47 ANSWERS

A1 Genes on a particular chromosome normally will be inherited together, as a linkage group, unless they are separated during crossing-over. Genes located on the sex chromosomes normally will be inherited along with sex (male or female) and are said to be sex-linked. Genes located on the other chromosomes (autosomes) are said to show autosomal linkage.

A2 Approximately equal numbers of offspring with the genotypes $X^G X^G$, $X^G X^B$, $X^G Y$ and $X^B Y$ — a ratio of 1 ginger female : 1 tortoiseshell female : 1 ginger male : 1 black male

A3 Approximately equal numbers of offspring with the genotypes $X^H X^H$, $X^H X^h$, $X^H Y$ and $X^h Y$ — a ratio of 1 normal female : 1 carrier female : 1 normal male : 1 haemophiliac male.

***examiner's* note** It is important to indicate both sex *and* genotype when doing genetic crosses involving sex-linkage. Only females can be carriers — they possess an allele that is not expressed in the phenotype but which can be passed on.

A2 Biology
Genetics and evolution

48

Genetic crosses: techniques

The sequence of reactions involved in producing green eye colouration in cats is summarised below:

substrate (colourless) $\xrightarrow{A}$ intermediate (blue) $\xrightarrow{B}$ product (green)

If the dominant allele A is present, a cat can convert the substrate into the intermediate substance. If the dominant allele B is present, a cat can convert the intermediate substance into the final product.

Q1 What would be the colour of a cat's eyes if only A was present?

Q2 What would be the colour if both A and B were present?

48 ANSWERS

A1 Blue

A2 Green

examiner's note The key things to remember when doing genetic crosses are:
- set your work out clearly, labelling the parents, gametes and new generations
- do not confuse the terms 'genotype' and 'phenotype', or 'gene' and 'allele'
- remember that gametes are haploid, so they only have one of each pair of alleles
- if you do not find the ratio you expect, it may be a case of linkage
- when dealing with a sex-linked characteristic, make sure that you indicate both the sex and the genotype of each individual

A2 Biology
Genetics and evolution

49

Genetic crosses: analysing

Q1 Are gametes diploid or haploid?

Q2 Why is it rare that a genetic cross will produce the exact genetic ratio expected?

Q3 A cross between two heterozygous tall pea plants with purple flowers gives the ratio of phenotypes below. Are these results significantly different from the expected ratio of 9:3:3:1?

| 152: | 38: | 41: | 24 |
| tall purple | tall white | short purple | short white |

ANSWERS

49 ANSWERS

A1 Haploid

A2 Because the number of offspring may be too small and chance factors will influence the actual numbers of each type of offspring produced

A3 To determine this a χ^2 (chi-squared) test is used. The calculated value of (χ^2) is 8.6. This is greater than the critical value (7.82) and so the null hypothesis is rejected. The results are not due to chance alone, i.e. the observed results are significantly different from the expected ratio.

***examiner's* note** In order to determine whether an unusual ratio is significantly different from that expected, e.g. due to linkage, biologists use a chi-squared test. This test shows whether the difference between observed and expected results is big enough to be *significant*, or whether it is just due to *chance*.

A2 **Biology**
Genetics and evolution

Variation

Q1 (a) What is continuous variation?
(b) Give an example of continuous variation in humans.

Q2 (a) What is discontinuous variation?
(b) Give an example of discontinuous variation in humans.

Q3 Distinguish between continuous and discontinuous variation in terms of:
(a) the number of genes involved
(b) the influence of the environment

ANSWERS

ANSWERS

A1 (a) A variation of a characteristic within a population such that a complete range of forms can be seen
(b) Height or weight

A2 (a) A type of variation in which there are clearly defined differences within a population
(b) ABO blood groups

A3 (a) There are many genes (polygenes) involved in continuous variation, but only a few (often one or two) involved in discontinuous variation
(b) The environment may have a significant effect on continuous variation but usually has little or no effect on discontinuous variation

***examiner's* note** It is important to be able to distinguish between continuous and discontinuous variation. The key differences are:
- the number of genes involved
- the influence of the environment
- the typical frequency distribution of phenotypes

A2 Biology
Genetics and evolution

Meiosis

Q1 What is meiosis?

Q2 State during which division of meiosis (first or second) the following occur:
(a) chromatids separate
(b) homologous chromosomes form pairs
(c) chiasmata are formed
(d) independent assortment of chromosomes occurs

Q3 Explain how meiosis can lead to genetic variation.

51 ANSWERS

A1 Meiosis is a type of cell division which occurs in the production of gametes. It involves two successive divisions of a diploid cell to produce haploid gametes.

A2 (a) Second (b) First
  (c) First (d) First

A3 During meiosis, the homologous chromosomes exchange genetic material by crossing over at chiasmata. The chromosomes also show independent assortment. This means that gametes are genetically different from each other and so there will be genetic variation in the next generation.

examiner's **note** Genetic variation arises during meiosis as a result of crossing-over in prophase I, together with the random assortment of homologous chromosomes in metaphase I and random fertilisation during sexual reproduction.

A2 Biology
Genetics and evolution

52

Mutation

Q1 Explain what is meant by the term mutation.

Q2 Distinguish between point and chromosome mutations.

Q3 Consider a single mutation in a gene coding for a metabolic protein. Explain how it might be:
(a) advantageous
(b) harmful
(c) neutral

ANSWERS

ANSWERS

A1 A change in the amount or arrangement of DNA in a cell

A2 A point (gene) mutation is any change in the sequence of bases in DNA. A chromosome mutation is any change in the structure or number of chromosomes.

A3 The change in the amino acid sequence may change the shape of the protein. This may cause the metabolic protein to function:

(a) more efficiently giving an organism a selective *advantage* in the environment

(b) less efficiently giving an organism a selective *disadvantage* in the environment

(c) as before, in which case the mutation will be *neutral* and offer no selective advantage or disadvantage

***examiner's* note** Most mutations are harmful and cause genetic diseases. However, mutations can occasionally provide an individual with an advantage over competitors and they are an important source of new inherited variation in evolution.

A2 Biology
Genetics and evolution

Polysomy and polyploidy

Q1 Distinguish between polysomy and polyploidy.

Q2 Distinguish between autopolyploidy and allopolyploidy.

Q3 Complete the diagram below showing the relationship between the different types of mutation of chromosome number.

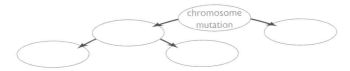

ANSWERS

53 ANSWERS

A1 Polysomy is a condition in which an organism has at least one more chromosome than normal — the number of a particular chromosome is not diploid. Polyploidy is a condition in which an organism has three or more sets of chromosomes in each cell.

A2 Autopolyploidy or 'self-polyploidy' occurs when chromosomes fail to separate during cell division — all the chromosomes are from the same species. Allopolyploidy or 'cross-polyploidy' involves hybridisation between two different species — the resulting hybrid is usually sterile.

A3

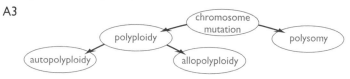

***examiner's* note** Most cases of polysomy are lethal in animals, except those involving the smaller autosomes or sex chromosomes. Polyploidy is very rare in animals.

A2 Biology
Genetics and evolution

54

Population genetics

Q1 What is a gene pool?

Q2 A characteristic is controlled by one gene with two alleles, B and b. The frequency of allele B in the population is 0.6 (p).
(a) What is the frequency of allele b in the population (q)?
(b) What is the frequency of the heterozygous genotype (Bb) in the population?

ANSWERS

ANSWERS

A1 All the genes or alleles present in a population (group of organisms of the same species living in a particular area)

A2 (a) Frequency of allele b = $(1 - p) = (1 - 0.6) = 0.4$ (q)
  (b) Frequency of Bb genotype = $2pq = 2 \times 0.6 \times 0.4 = 0.48$ (48%)

***examiner's* note** The Hardy–Weinberg principle that the gene frequencies in a population remain constant from generation to generation only applies in certain conditions:
- a large population
- random mating
- no migration into or out of the population
- no mutations or natural selection

If any of these conditions are not met, allele frequencies will start to change — this is evolution.

A2 Biology
Genetics and evolution

Evolution

Q1 What is evolution?

Q2 What type of variation is important in evolution?

Q3 Using Darwin's theory of evolution, explain how the long necks of giraffes may have evolved.

ANSWERS

55 ANSWERS

A1 The process by which new species arise as the result of gradual change to the genetic make-up of existing species over long periods of time

A2 Genetic variation

A3
- An ancestral population of giraffes would have shown genetic variation in neck length.
- There would have been competition for resources and giraffes with longer necks would have had a selective advantage over those with shorter necks as they could feed on higher tree vegetation
- Giraffes with longer necks were more likely to survive and reproduce, passing on the genes for longer necks

***examiner's* note** A common example is the evolution of resistance to antibiotics in bacteria. Antibiotics do *not* cause the bacteria to mutate. The mutation (gene for resistance) already exists and the use of antibiotics simply selects those bacteria that carry this mutation. The resistant bacteria then reproduce and, over a period of time, the whole population becomes resistant.

A2 Biology
Genetics and evolution

56

Directional selection

Q1 What is directional selection?

Q2 Give an example of directional selection.

Q3 Insert the missing word: 'selection affects ………… frequencies'.

ANSWERS

ANSWERS

A1 A type of natural selection which occurs when environmental change favours a new form (phenotype) of an organism

A2 In soils contaminated with heavy metals, such as lead, tin, or zinc, natural selection will favour grasses that can tolerate these metals

A3 Allele

***examiner's* note** Selection affects allele frequencies. Directional selection is a type of natural selection in which individuals of one extreme type are more likely to survive and reproduce. This occurs when the environment changes and favours a particular type of individual.

A2 Biology
Genetics and evolution

Disruptive selection

Q1 What is disruptive selection?

Q2 Give an example of disruptive selection.

Q3 Identify the types of natural selection described below:
(a) A type of natural selection in which the environment favours forms at the extremes of the range of phenotypic variation.
(b) A type of natural selection which occurs when environmental change favours a new form (phenotype) of an organism.

ANSWERS

A1 A type of natural selection in which the environment favours forms at the extremes of the range of phenotypic variation

A2 Disruptive selection may have occurred during the evolution of specialised gametes (eggs and sperm). In this case, small gametes (sperm) could be produced in large numbers, and large gametes (eggs) would have sufficient energy to survive and nourish the initial growth of the zygote. Both would have advantages over medium-sized gametes, which would be selected against.

A3 (a) Disruptive selection
(b) Directional selection

***examiner's* note** Disruptive selection occurs when selection favours the extreme trait values over the intermediate trait values. In this case, the variance increases as the population is divided into two distinct groups. Disruptive selection plays an important role in speciation.

A2 Biology
Genetics and evolution

58

Stabilising selection

Q1 What is stabilising selection?

Q2 Give an example of stabilising selection.

Q3 Tiger sharks have changed very little in form over the last 10 million years. What does this suggest about:
(a) the genetic make-up of tiger sharks?
(b) how well tiger sharks are adapted to their environment?
(c) the nature of the tiger sharks' environment?

ANSWERS

ANSWERS

A1 A type of natural selection in which the environment acts against forms at the extremes of the range of phenotypic variation

A2 Babies whose birth weight is significantly below or above the mean of 3.6 kg have greater mortality than babies of average birth weight

A3 (a) There is little genetic variation between tiger sharks — the gene pool is stable
(b) They are well adapted to their environment — there has been little selection pressure for change
(c) The environment of the tiger sharks is relatively stable — there has been little change in abiotic or biotic factors

***examiner's* note** Stabilising selection is where individuals with traits towards the middle of the range are more likely to survive and reproduce. It is the most common type of natural selection and occurs when the environment is not changing significantly.

A2 Biology
Genetics and evolution

Adaptations to the environment

Q1 Describe how animals may be structurally adapted to living in cold climates.

Q2 Outline a physiological adaptation seen in desert mammals.

Q3 Describe how cacti are adapted to living in arid environments.

59 ANSWERS

A1 Penguins have a small surface area to volume ratio and are structurally adapted to living in cold environments by having a thick layer of insulating fat and waterproof insulating feathers

A2 The kidneys of desert mammals are adapted in such a way as to produce very concentrated urine, thereby saving water

A3 Cacti have extensive root systems to take up water, various mechanisms to reduce water loss by transpiration, as well as water storage systems to survive arid environments

***examiner's* note** There are three major types of adaptation:
- size and shape, e.g. large ears in elephants to lose heat
- specialised features, e.g. the ability of halophytes to live in conditions of high salinity
- behavioural adaptations, e.g. bird migration

A2 Biology
Genetics and evolution

60

Speciation: allopatric

Q1 Explain what is meant by the term speciation.

Q2 Describe and explain the conditions essential for speciation to occur.

Q3 Describe what is meant by the term allopatric speciation.

ANSWERS

ANSWERS

A1 Speciation refers to the formation of a new species of organisms.

A2 Two populations of the parent species must become geographically or reproductively isolated so that gene flow between them is prevented. Interbreeding between the populations to produce fertile offspring eventually becomes impossible — the two populations have become different species.

A3 The development of one or more species, which occurs when populations of the parent species become geographically isolated from each other

***examiner's* note** The finches of the Galapagos Islands are a good example of allopatric speciation. Although the finches had a common ancestor, geographical isolation has resulted in the formation of 14 different species of finch. Each species of finch inhabits a different ecological niche on the islands and is adapted to local environmental conditions.

A2 Biology
Genetics and evolution

Speciation: sympatric

Q1 Describe what is meant by the term sympatric speciation.

Q2 Name the four main causes of reproductive isolation.

Q3 A male donkey and a female horse can reproduce successfully to produce a viable offspring (a mule). However, this offspring is infertile. What kind of isolating mechanism separates donkeys and horses?

ANSWERS

61 ANSWERS

A1 Sympatric speciation is the development of one or more species, which occurs when populations of the parent species living in the same area are prevented from interbreeding — the two populations are said to be reproductively isolated from each other.

A2
- Seasonal isolation
- Mechanical isolation
- Behavioural isolation
- Gametic isolation

A3 Post-zygotic isolating mechanism

examiner's note Most speciation is *intraspecific* — new species arise from a common ancestor due to allopatric or sympatric speciation, or a combination of the two processes. *Interspecific* speciation (when a new species arises due to the combination of two different parent species) can also occur, but is rare in animals. It can be seen in plants as a result of allopolyploidy.

A2 Biology
Classification

Taxonomy

Q1 A population of birds becomes geographically separated into two groups. The two groups eventually become separate species. The two new species of bird are members of the same genus and the same kingdom.
(a) Name three other taxonomic groups to which both species of birds would belong.
(b) Name the kingdom to which both species belong.

Q2 For humans, name the following:
(a) phylum   (b) class   (c) order

ANSWERS

ANSWERS

A1 (a) Any three of:
- Phylum
- Order
- Class
- Family

(b) Animalia

A2 (a) Chordata
(b) Mammalia
(c) Primates

***examiner's* note** Cladistics is a method of taxonomy that focuses on the features of organisms that are evolutionary developments. This emphasis on phylogeny (the genetic relationship between organisms) is useful as it enables us to determine points in time when one species split into two, such as when humans and chimpanzees split from a common ancestor around 6 million years ago.

A2 Biology
Classification

63

Animals and fungi

Q1 What are the common features shared by all animals?

Q2 What are the common features shared by all fungi?

Q3 Give two examples of fungi.

ANSWERS

63 ANSWERS

A1
- They are multicellular organisms
- Their cells do not possess cell walls
- They show heterotrophic nutrition

A2
- They are heterotrophic
- They reproduce by means of spores
- They are usually made up of thread-like structures known as hyphae
- Most have cell walls made of chitin

A3 Any two of:
- Mushrooms
- Toadstools
- Moulds
- Yeast

***examiner's* note** Animals and fungi are eukaryotic, multicellular and heterotrophic. The key difference between these kingdoms at a cellular level is that fungus cells have cell walls (made of chitin) and animal cells do not.

A2 Biology
Classification

64

Plants, protoctista and prokaryotes

Q1 What are the common features shared by all plants?

Q2 What are the distinguishing features of the kingdom Prokaryotae?

Q3 (a) What are protoctista?
(b) Give one example of an organism belonging to the kingdom Protoctista.

ANSWERS

ANSWERS

A1
- They are multicellular organisms
- Their cells possess cell walls composed mainly of cellulose
- They show autotrophic nutrition

A2 Prokaryotae is the kingdom that includes bacteria. These organisms have a cell wall. Prokaryotic cells have a different structure to all other (eukaryotic) cells — they are usually smaller than eukaryotic cells, they have no true nucleus, they have no membrane-bound organelles.

A3 (a) Protoctista is the kingdom containing eukaryotic organisms that cannot be classified as members of the other kingdoms
(b) Red algae, slime moulds or amoeba

***examiner's* note** Plants and protoctista are eukaryotic. Protoctista is an unusual kingdom in that it contains all the species that do not fit into the other eukaryotic kingdoms. Prokaryotes are unicellular and differ from all the other kingdoms due to their cellular structure.

A2 Biology
Homeostasis

Homeostasis

Q1 Explain what is meant by the term homeostasis.

Q2 Distinguish between the terms positive feedback and negative feedback.

Q3 Which important factors must be kept constant in intercellular fluid?

ANSWERS

ANSWERS

A1 The maintenance of a constant internal environment by using physiological mechanisms to regulate important factors, such as body temperature (thermoregulation)

A2 Positive feedback is a condition in which the deviation of a physiological factor from its optimum causes further deviation to occur. Negative feedback is a condition in which the deviation of a physiological factor from its optimum initiates mechanisms that are used to return the factor to its optimum.

A3 It must contain the correct quantities of oxygen and nutrients, and have the optimum pH, temperature and water potential

examiner's **note** Examples of homeostasis include thermoregulation (the control of body temperature in animals), the control of respiratory gases in the blood and the regulation of blood glucose. It is important to understand that homeostasis allows organisms to be independent of the external environment. Therefore, animals with efficient homeostatic mechanisms (e.g. mammals) can survive in a wide range of habitats.

A2 Biology
Homeostasis

The regulation of blood glucose I

Q1 Why is glucose a vital nutrient in all living cells?

Q2 Distinguish between hyperglycaemia and hypoglycaemia.

Q3 Distinguish between insulin dependent diabetes and non-insulin dependent diabetes.

ANSWERS

A1 It is the main substrate for respiration

A2 Hyperglycaemia is an abnormally high level of glucose in the blood, resulting in dehydration and weight loss. Hypoglycaemia is an abnormally low level of glucose in the blood, resulting in tiredness, sweating and possibly coma.

A3 In insulin dependent diabetes, the pancreas fails to produce enough insulin to control the concentration of glucose in the blood. This condition can be controlled by regulating the diet and injecting insulin.

In non-insulin dependent diabetes, the pancreas produces enough insulin but the cells of the body fail to take up glucose in response to the hormone. It is mainly controlled by regulating the diet.

***examiner's* note** Type I diabetes (insulin dependent) is often diagnosed in childhood. Type II diabetes (non-insulin dependent), however, usually develops in middle age. Type II diabetes is associated with obesity or a poor quality diet.

A2 Biology
Homeostasis

The regulation of blood glucose II

Q1 Explain why the blood concentration of glucagon rises during exercise, while that of insulin falls.

Q2 Explain how one molecule of glucagon can bring about the conversion of many molecules of glycogen to glucose.

Q3 Bob has diabetes and is experiencing high concentrations of glucose in his blood. He is injected with insulin but this has little effect on the high glucose concentrations. What type of diabetes does Bob have?

ANSWERS

67 ANSWERS

A1 During exercise, the skeletal muscles rapidly respire and therefore use a lot of glucose. Homeostatic mechanisms are required to increase the level of glucose in the blood. Since glucagon stimulates a rise in blood glucose concentration, this hormone is secreted by the pancreas. Insulin (which stimulates the concentration of blood glucose to fall) is not secreted.

A2 Many hormones operate on a cascade principle. When one molecule of glucagon binds to its receptor, it activates an enzyme which catalyses the breakdown of glycogen to glucose. One enzyme molecule can catalyse the breakdown of many glycogen molecules. The enzyme is said to have a high turnover rate.

A3 Non-insulin dependent diabetes

examiner's note Blood glucose concentration increases after consuming food, especially if it is high in carbohydrate. The glucose concentration falls after exercise, since more glucose is needed for respiration to release energy.

A2 Biology
Homeostasis

68

The regulation of blood glucose III

Q1 Distinguish between gluconeogenesis and glycogenolysis.

Q2 Explain why insulin and glucagon affect mainly skeletal muscle cells and liver cells.

Q3 Name one hormone, other than insulin and glucagon, that can influence the concentration of plasma glucose.

ANSWERS

A1 Gluconeogenesis is the synthesis of glucose from non-carbohydrate sources, such as lipid and protein. Glycogenolysis is the conversion of glycogen to glucose.

A2 Insulin and glucagon bind to protein receptors in the plasma membranes of skeletal muscle and liver cells. The protein receptors have a specific shape so that only these hormones will fit. Other body cells do not have these specific receptors.

A3 Adrenaline

***examiner's* note** Insulin reduces blood glucose in three ways:
- increasing uptake of glucose by liver cells and muscle cells
- stimulating glycogenesis in these cells
- increasing the rate of respiration

Glucagon increases blood glucose by stimulating glycogenolysis.

A2 Biology
Homeostasis

Thermoregulation

Q1 What is thermoregulation?

Q2 Distinguish between the terms endothermic and ectothermic.

Q3 List three heat retention mechanisms.

ANSWERS

ANSWERS

A1 The process of maintaining a relatively constant body temperature, despite fluctuations in the temperature of the environment

A2 Endothermic refers to animals such as mammals and birds that use mainly internal (physiological) mechanisms to control temperature. Ectothermic refers to animals that use mainly external (behavioural) mechanisms to control temperature.

A3 Any three of:
- Increased metabolic rate
- Shivering
- Vasoconstriction (decreased blood flow to skin)
- Body hairs raised to increase insulating air layer

***examiner's* note** Animals are either ectothermic or endothermic. Endothermic animals generally have more control over their body temperature.

A2 Biology
Homeostasis

70

The liver: metabolism

Q1 Describe the role of the liver in the metabolism of lipids.

Q2 Describe the role of the liver in the removal of harmful substances from the body.

Q3 Apart from the processes mentioned in Q1 and Q2, name two other roles of the liver in homeostasis.

ANSWERS

ANSWERS

A1 Excess fatty acids and glycerol are converted into triglycerides and these are transported to storage sites as lipoproteins. Triglycerides can be hydrolysed by lipase enzymes to release fatty acids and glycerol for respiration. The liver also helps to regulate the cholesterol content of the blood.

A2 Alcohol, drugs and other toxins are converted into less harmful compounds by the liver so that they can be excreted from the body

A3 Any two of:
- Regulation of blood glucose
- Production of bile
- Synthesis of plasma proteins

examiner's note The liver performs a wide range of metabolic functions, which are important for homeostasis and the production of excretory products for removal from the body.

A2 Biology
Homeostasis

The liver: excretion

Q1 What is deamination?

Q2 Name the following from the ornithine cycle:
(a) the substrates
(b) the product

Q3 Name the process by which the liver converts one amino acid into another.

ANSWERS

71 ANSWERS

A1 The removal of nitrogen-containing amino groups from excess amino acids, forming ammonia and organic acids. The organic acids are respired or converted to carbohydrate and stored as glycogen.

A2 (a) Ammonia and carbon dioxide
(b) Urea

A3 Transamination

***examiner's* note** Mammal livers produce urea, but non-mammals produce different types of nitrogenous waste. For example, to use less energy, freshwater fish produce ammonia — the surrounding excess water dilutes this toxic compound. To conserve water, birds and insects produce semi-solid uric acid.

A2 **Biology**
Homeostasis

The mammalian kidney I

Q1 What is osmoregulation?

Q2 Put these parts of a kidney nephron in the correct order:
- proximal convoluted tubule
- collecting duct
- distal convoluted tubule
- Bowman's capsule
- loop of Henle

Q3 What are the three main processes involved in excretion?

A1 The maintenance of the optimum water potential of the blood

A2
- Bowman's capsule
- proximal convoluted tubule
- loop of Henle
- distal convoluted tubule
- collecting duct

A3
- Ultrafiltration
- Selective reabsorption
- Removal of nitrogenous waste (urine)

***examiner's* note** Excretion is the removal of the waste products of cellular metabolism. Defaecation (the removal of faeces) is *not* the same as excretion, because the waste materials are not the products of cellular metabolism. Similarly, it is important not to confuse excretion with secretion, which is the transport of *useful* substances to their place of action.

A2 **Biology**
Homeostasis

The mammalian kidney II

Q1 Explain whether the reabsorption of all the glucose in glomerular filtrate back into the blood is due to diffusion, active transport or osmosis.

Q2 Glucose is reabsorbed in the proximal convoluted tubule. Give two features of the cells lining this tubule that make them well adapted for reabsorption.

Q3 Will a longer loop of Henle cause the production of *more* or *less* concentrated urine?

ANSWERS

A1 Some reabsorption is due to diffusion, as the concentration of glucose in glomerular filtrate is initially higher than that in the blood plasma. However, some reabsorption is due to active transport, otherwise it would not be possible to reabsorb all the glucose. Osmosis is not involved as only water moves by osmosis.

A2
- The presence of microvilli on the surface of the cells increases the surface area available for the reabsorption of glucose
- They have many mitochondria, to supply the ATP required for the active transport of glucose

A3 More concentrated

***examiner's* note** Glomerular filtrate contains many useful substances filtered out of the blood, such as glucose. These substances have to be reabsorbed from the nephron back into the blood. Glomerular filtrate is therefore *not* the same as urine.

A2 Biology
Homeostasis

74

The mammalian kidney III

Q1 What process is responsible for the removal of sodium and chloride ions from the ascending limb of the loop of Henle?

Q2 A solution with a high solute concentration has a very negative water potential. True or false?

Q3 What causes the increase in urea concentration as filtrate passes through the collecting duct?

ANSWERS

A1 Active transport

A2 True

A3 Water is reabsorbed from glomerular filtrate as it passes through the nephron, so by the time the filtrate reaches the collecting duct the concentration of urea is much higher than in the blood plasma. As the filtrate passes down the collecting duct, more water is reabsorbed back into the blood, increasing the concentration of urea even further.

***examiner's* note** Note that a longer loop of Henle will result in the production of more concentrated urine. Therefore animals that live in arid regions, such as the kangaroo rat, have relatively long loops of Henle to conserve water; those living in water-rich regions, such as the beaver, have shorter loops of Henle.

A2 **Biology**
Homeostasis

The mammalian kidney IV

Q1 Where in the human body are the osmoreceptors located?

Q2 Name the gland that produces antidiuretic hormone (ADH).

Q3 Animals living in water-rich environments tend to have relatively short loops of Henle and low concentrations of ADH in their blood. Suggest an explanation for this.

ANSWERS

A1 In the hypothalamus in the brain

A2 Pituitary gland

A3 Movement of ions across the walls of the loop of Henle enables it to act as a countercurrent multiplier, building up the concentration of sodium in the medulla of the kidney. This process results in the reabsorption of water by osmosis and the production of concentrated urine in the collecting duct. Short loops of Henle will result in less concentrated urine. Low concentrations of ADH will mean that the collecting ducts are less permeable to water and so not much water is reabsorbed back into the blood from the kidney nephrons. Little water will be conserved, but this is not relevant to animals with easy access to water.

examiner's **note** Aldosterone is another hormone involved in osmoregulation. It regulates the concentration of sodium ions in the blood.

A2 Biology
Coordination in plants and mammals

Plant growth substances I

Q1 Name a plant growth substance that:
(a) stimulates cell growth (b) inhibits germination in seeds

Q2 Name a plant growth substance that:
(a) promotes the growth of lateral buds and leaves in stems
(b) inhibits root growth at high concentrations

Q3 List the ways in which the following substances influence physiological processes in plants:
(a) abscisic acid (b) cytokinins

ANSWERS

ANSWERS

A1 (a) Auxin
(b) Abscisic acid

A2 (a) Cytokinin
(b) Auxin

A3 (a) Stimulates the closing of stomata when water is in short supply, may play a part in leaf fall in some deciduous trees and inhibits germination in seeds
(b) Stimulate cell division, promote the growth of lateral buds and leaves in stems and delay senescence (ageing) in leaves

***examiner's* note** In plants, coordination and response takes place by chemical communication between cells. It takes time for the chemicals to reach their targets, so plant response times are slower than those of animals — abscisic acid inhibits growth, auxins stimulate cell growth and cytokinins stimulate cell division.

A2 Biology
Coordination in plants and mammals

Plant growth substances II

Q1 Name a plant growth substance that stimulates:
(a) the ripening of fruit
(b) germination

Q2 Distinguish between synergistic and antagonistic interaction among plant growth substances.

Q3 State four commercial uses for plant growth substances, identifying the substance(s) involved in each case.

ANSWERS

ANSWERS

A1 (a) Ethene
(b) Gibberellin

A2 When plant growth substances act synergistically, their combined action produces a greater effect than would be expected from adding the individual effects of each substance. Conversely, when they act antagonistically, their combined action produces a smaller effect than would be expected from adding the individual effects of each substance.

A3
- Promoting the rooting of cuttings — auxins
- Selective killing of weeds — auxins
- Inducing fruit ripening — ethene
- Producing seedless fruit — gibberellins

***examiner's* note** Auxins and gibberellins are synergistic growth regulators. Abscisic acid and ethene are antagonistic to auxins and gibberellins.

A2 Biology
Coordination in plants and mammals

The phytochrome system

Q1 What are photoreceptors?

Q2 At the end of a long summer's day, a lot of phytochrome P_R has been changed into phytochrome P_{FR}. True or false?

Q3 Explain the importance of light in phototropism

ANSWERS

78 ANSWERS

A1 Cells that are sensitive to light

A2 True

A3 Phototropism is a change in growth in response to light. Most stems are positively phototropic because they grow towards light. Roots, on the other hand, are negatively phototropic, growing away from light.

***examiner's* note** When light intensity or wavelength determines the timing of a biological process, it is known as photoperiodism. The flowering time of many plants is determined by the length of the day (the photoperiod) and controlled by phytochromes. Different species react in different ways to photoperiods:
- Long-day plants – flower in the summer.
- Short-day plants – flower in the winter.
- Day-neutral plants – flower independently of day length.

A2 Biology
Coordination in plants and mammals

Non-sex hormones

Q1 Insert the missing words: 'Hormones bind to .... on the cell surface membrane of .... cells, where they bring about a response.'

Q2 Name the hormone that (a) stimulates a reduction in the concentration of glucose in the blood, (b) inhibits the production of dilute urine and (c) stimulates an increase in the concentration of glucose in the blood.

Q3 Distinguish between the effects of peptide and steroid hormones on a target cell.

ANSWERS

ANSWERS

A1 receptors, target

A2 (a) Insulin
(b) Antidiuretic hormone (ADH)
(c) Glucagon

A3 Peptide hormones attach to a receptor on the cell surface membrane and activate adenyl cyclase. This converts ATP to cAMP. The cAMP then activates an enzyme system within the cell. Steroid hormones pass through the cell surface membrane and form a complex with a receptor molecule. This complex passes into the nucleus and increases transcription, resulting in increased protein synthesis.

***examiner's* note** Many types of hormones are proteins or smaller peptides, e.g. insulin and adrenaline. Other hormones are steroids, e.g. oestrogen, progesterone and testosterone.

A2 Biology
Coordination in plants and mammals

80

Sex hormones

Q1 Name the hormone that (a) stimulates the development of male secondary sexual characteristics and (b) maintains the lining of the uterus.

Q2 Name the hormone that stimulates (a) the development of female secondary sexual characteristics, (b) ovulation and (c) the production of gametes.

Q3 Explain why progesterone is used in oral contraceptives.

ANSWERS

ANSWERS

A1 (a) Testosterone
(b) Progesterone

A2 (a) Oestrogen
(b) Luteinising hormone (LH)
(c) Follicle-stimulating hormone (FSH)

A3 Progesterone inhibits the production of FSH and so prevents development of an ovum. Progesterone also inhibits the production of LH and so prevents ovulation.

***examiner's* note** The stages of the menstrual cycle are controlled by hormones and monitored by negative feedback loops to keep the system in balance. For example, FSH and LH stimulate the production of oestrogen and progesterone. When oestrogen and progesterone concentrations get sufficiently high, they inhibit the release of FSH and LH. Consequently, any further production of oestrogen and progesterone is also inhibited.

A2 Biology
Coordination in plants and mammals

81

Nervous and chemical coordination

Q1 State three differences between nervous and hormonal coordination.

Q2 How are the nervous and hormonal systems linked in the brain?

Q3 Name a hormone synthesised by neurosecretory cells in the hypothalamus.

ANSWERS

A1
- Nervous coordination is faster than hormonal coordination
- Nervous coordination is more short-lived than hormonal coordination
- Nervous coordination involves electrical information moving along neurones, whereas hormonal coordination involves chemical information moving via the circulatory system

A2 The hypothalamus is a small part of the brain that links the nervous and endocrine systems. It controls hormone secretion via the pituitary gland located below it. The hormones secreted by the pituitary gland control hormone secretion by other endocrine glands.

A3 Antidiuretic hormone (ADH) or oxytocin

examiner's note There are several key differences between nervous and hormonal coordination. Nevertheless, nerves and hormones have to work together in order to coordinate internal body functions and appropriate responses to the external environment.

A2 Biology
The nervous system

Neurones

Q1 What is a neurone?

Q2 What are the functions of the myelin sheath?

Q3 Identify the types of neurones that do the following:
(a) transmit impulses from the central nervous system to an effector
(b) connect sensory and motor neurones within the central nervous system
(c) transmit impulses from a receptor to the central nervous system

A1 A cell specialised for the conduction of nerve impulses

A2
- To insulate and protect the neurone
- To speed up the conduction of nerve impulses

A3 (a) Motor neurones
(b) Intermediate neurones
(c) Sensory neurones

examiner's note Neurones carry waves of electrical activity called action potentials (nerve impulses). They can do this since their cell membranes are polarised — there are different charges on the inside and outside of the membrane.

A2 Biology
The nervous system

83

Receptors and effectors

Q1 (a) What are receptors?
(b) Outline how a stimulus might trigger a response in a sensory neurone.

Q2 Name three types of receptor found in humans.

Q3 (a) What are effectors?
(b) State two types of effector found in animals.

ANSWERS

ANSWERS

A1 (a) Cells that are specialised to detect a particular stimulus
(b) If the stimulus exceeds a minimum threshold value, the receptor is depolarised and an action potential is set up in a sensory neurone

A2 Any three of:
- Baroreceptors (blood pressure)
- Chemoreceptors (blood pH)
- Photoreceptors (light)
- Mechanoreceptors (sound/pressure)

A3 (a) Cells or organs that respond to a stimulus
(b) Muscles and glands

***examiner's* note** Organisms have receptors that are sensitive to stimuli. Receptor cells have excitable membranes which enable them to receive a stimulus and set in motion a response to this stimulus. An effector is a part of the body that brings about a response to the signal from a receptor. In animals, effectors are usually muscles (the response is contraction) or glands (the response is secretion).

A2 Biology
The nervous system

The nerve impulse I

Q1 What is meant by the term resting potential?

Q2 What is a typical resting potential of a human neurone?

Q3 Describe the changes that occur to the restoration of the resting potential when a neurone is stimulated.

ANSWERS

A1 In an unstimulated neurone, there is a high concentration of sodium ions outside the cell. The inside is negative relative to the outside.

A2 Around −60 mV

A3 Stimulation of the neurone causes sodium ions to enter the cell, making it positive relative to the outside. This is known as depolarisation. Potassium ions then leave the cell, restoring the negative potential difference — repolarisation. The true resting potential is then restored by sodium–potassium pumps, which actively transport potassium into the cell and sodium out of the cell.

***examiner's* note** The rapid change in electrical charge across the membrane of a nerve cell causes the transmission of an impulse and is known as an action potential. In other words, stimulation of a nerve cell causes the electrical charge across the membrane to change from the resting potential of −60 mV to +40 mV. This change in electrical charge, which lasts for only a few milliseconds before the resting potential is restored, is the action potential.

A2 Biology
The nervous system

The nerve impulse II

Q1 What is meant by the term all-or-nothing?

Q2 What is the refractory period of a neurone?

Q3 Name the three factors that speed up the transmission of nerve impulses along axons.

ANSWERS

85 ANSWERS

A1 All-or-nothing is a term used to describe the fact that action potentials in nerve cells are always identical in size.

A2 The time interval during which a nerve cell is incapable of responding to a stimulus.

A3
- Diameter of the axon — impulses travel faster in larger axons because there is less electrical resistance
- Temperature — impulses go faster as temperature increases, up to around 40°C (above this temperature, the proteins begin to denature)
- Presence of a myelin sheath

***examiner's* note** Nerve axons in vertebrates are usually covered in a myelin sheath, which is an electrical insulator. Between the sheaths there are small patches of bare membrane called nodes of Ranvier, where sodium channels are concentrated. Impulses jump from one node to the next, which lets them move faster and is known as saltatory conduction.

A2 Biology
The nervous system

The synapse I

Q1 What is a synapse?

Q2 Name the two most common neurotransmitters found in synapses.

Q3 Insert the missing words: (a) 'Vesicles containing neurotransmitter fuse with the .... and release their contents into the synaptic cleft.' (b) 'Neurotransmitter diffuses across the cleft and attaches to .... on the post-synaptic membrane.'

ANSWERS

A1 The junction between two nerve cells

A2 Acetylcholine and noradrenaline

A3 (a) pre-synaptic membrane
(b) receptor molecules

***examiner's* note** Neurotransmitter receptors are present only on the post-synaptic membrane, so a signal can only pass across a synapse in one direction. The post-synaptic cell behaves as a transducer, just like receptor cells, because the chemical stimulus (neurotransmitter) is converted into an electrical one (action potential).

A2 Biology
The nervous system

The synapse II

Q1 How does an inhibitory neurotransmitter work?

Q2 What is a neuromuscular junction?

Q3 Distinguish between temporal and spatial summation.

ANSWERS

ANSWERS

A1 It makes the post-synaptic membrane hyperpolarised and therefore harder to excite

A2 The point at which a motor neurone connects with a muscle

A3 Summation is the process that occurs in synapses due to the additive effect of a number of stimuli. A single stimulus may not be sufficient to trigger an action potential, usually because there is not enough neurotransmitter produced. In temporal summation, two or more impulses arriving at the synapse in a short period of time may generate an action potential — the effects of the impulses add up over time. In spatial summation, an action potential may be triggered if several synapses act at the same time on the post-synaptic cell.

***examiner's* note** The neurotransmitter acetylcholine binds to cholinergic receptors. Noradrenaline binds to adrenergic receptors.

A2 Biology
The nervous system

Muscles I

Q1 Name the three main types of muscle.

Q2 What is meant by the term myogenic?

Q3 Insert the missing words: 'Striated muscle is made up of muscle .... each of which is made up of a number of myofibrils. Each myofibril is made up of myofilaments. The pattern of .... myofilaments (made up of the protein ....) and .... myofilaments (made up of the protein ....) gives the striations in the muscle.'

ANSWERS

ANSWERS

A1
- Striated muscle
- Smooth muscle
- Cardiac muscle

A2 Muscle which contracts spontaneously without input from the nervous system

A3 fibres, thin, actin, thick, myosin

***examiner's* note** Muscle fibres are adapted to respond effectively to a stimulus from a motor neurone. The cytoplasm of the fibres (sarcoplasm) is penetrated by transverse tubules which make up a network called the sarcoplasmic reticulum. They allow the sarcolemma (cell membrane) to transmit action potentials in towards the myofilaments. There are also many mitochondria to provide the required ATP for contraction.

A2 **Biology**
The nervous system

Muscles II

Q1 Insert the missing words: 'Following stimulation, calcium ions bind to .... on the actin filaments. This causes another protein called .... to change position and unblock the binding sites on the actin filaments.'

Q2 'The myosin heads attach to the binding sites, forming .... bridges between the two filaments.'

Q3 '............ provides the energy for the cross-bridges to detach and reattach further along the actin filament.'

ANSWERS

A1 troponin
tropomyosin

A2 actomyosin

A3 ATP

***examiner's* note** When excitation of the muscle stops, calcium ions are actively pumped back out of the cytoplasm into the sarcoplasmic reticulum. This means that the troponin is released and the tropomyosin moves back to block the myosin binding site again. The cross-bridges are broken and the muscle fibre returns to its resting position.

A2 Biology
The nervous system

The eye: photoreceptors

Q1 What are photoreceptors?

Q2 Describe the features of rod cells according to the following criteria:
- (a) location
- (b) sensitivity to light
- (c) colour vision
- (d) visual acuity

Q3 Describe the features of cone cells according to the following criteria:
- (a) location
- (b) sensitivity to light
- (c) colour vision
- (d) visual acuity

ANSWERS

ANSWERS

A1 Cells that are sensitive to light — rod cells and cone cells

A2 (a) Throughout the retina (except in the fovea)
(b) High
(c) Absent
(d) Low

A3 (a) Concentrated in the fovea
(b) Low
(c) Present
(d) High

***examiner's* note** Cones give better visual acuity than rods because each cone connects with its own individual bipolar neurone, so it can send more detailed information to the brain (and provide colour vision). Rods are more sensitive to light than cones because many rods converge onto the same bipolar neurone. This means that the combined response of the rods can summate and be detected in the brain.

A2 Biology
The nervous system

The eye: visual pigments

Q1 Name the parts of the retina where there are:
(a) cones, but no rods
(b) neither cones nor rods

Q2 Name the visual pigment present in:
(a) rods
(b) cones

Q3 (a) Explain how we see a yellow colour.
(b) Explain why yellow objects appear grey in dim light.

ANSWERS

A1 (a) Fovea (b) Blind spot

A2 (a) Rhodopsin (b) Iodopsin

A3 (a) Colours are detected by cone cells. There are three types of cone, sensitive to red, green or blue light. Yellow light stimulates the red and green light receptors, but not the blue light receptors. Analysis of these impulses by the brain gives the perception of a yellow colour.

(b) Cone cells have a high light threshold, whereas rod cells are more sensitive to light. In dim light, only the rod cells send impulses to the brain. Rod cells have only one type of pigment (rhodopsin) causing images to be seen in black and white, or shades of grey in between.

***examiner's* note** Iodopsin has three different forms, which are maximally sensitive to either red, blue or green light. The relative activity of these red, blue and green cone cells determines the colour that is observed.

A2 Biology
The nervous system

The ear: balance

Q1 Name the three main sections of the ear.

Q2 Insert the missing words: 'Balance is regulated by the .... canals. There are three loops of canals containing a fluid called .... Each contains an .... Receptors in these structures detect fluid movement and send messages to the brain as the head changes position.'

Q3 Name the other ear cavities containing receptors that relay information on the position of the head and body.

ANSWERS

A1
- Outer ear
- Middle ear
- Inner ear

A2 semicircular, endolymph, ampulla

A3
- Utricle
- Saccule

***examiner's* note** The inner ear contains a system of three semicircular canals, the utricle and the saccule. These are linked to the middle canal of the cochlea. Ampullae in the semicircular canals detect direction and rate of movement of the head, and the utricle and saccule detect the position of the head relative to the body. Together these structures help to maintain balance.

A2 Biology
The nervous system

The ear: hearing

Q1 Insert the missing words: 'Sound waves are collected by the .... and channelled through the ear canal, making the ear drum vibrate.'

Q2 'Vibrations from the ear drum make the .... of the middle ear vibrate against the ....'

Q3 'Sensory hair cells on the membrane of the middle canal send impulses down the .... nerve to the brain for processing.'

93 ANSWERS

A1 pinna

A2 bones, oval window

A3 auditory

***examiner's* note** Different sounds produce waves of different frequencies and amplitudes. The pattern of the waves causes a corresponding pattern of excitation of sensory cells in the cochlea. The brain interprets this pattern as a distinct sound, or set of sounds.

A2 Biology
The nervous system

Reflexes

Q1 Explain what is meant by a spinal reflex.

Q2 Name the three types of neurone involved in a spinal reflex, in order of their stimulation.

Q3 A hand is placed on a hot object and immediately withdrawn. For this reflex action, name: (a) the stimulus, (b) the receptor, (c) the coordinator and (d) the effector.

ANSWERS

ANSWERS

A1 A rapid and automatic response to a stimulus that does not involve thought processes in the brain

A2
- Sensory neurone
- Relay or intermediate neurone
- Motor neurone

A3 (a) Touching the hot object
(b) Heat receptors in the hand
(c) Reflex arc in spinal cord
(d) Muscle(s) involved in the withdrawal of the hand

***examiner's* note** The simplest kind of reflex is a monosynaptic reflex, where the sensory neurone connects directly to the motor neurone. There is only one synapse in the arc within the central nervous system. An example is the knee jerk reflex. Action potentials do not pass to the brain and therefore no conscious thought is involved.

A2 Biology
The nervous system

The autonomic nervous system

Q1 What is the autonomic nervous system?

Q2 Describe the features of the sympathetic nervous system according to the following criteria: (a) neurotransmitter at synapses, (b) effect on heart rate and (c) effect on ventilation rate.

Q3 Describe the features of the parasympathetic nervous system according to the following criteria: (a) neurotransmitter at synapses, (b) effect on heart rate and (c) effect on ventilation rate.

ANSWERS

95 ANSWERS

A1 The part of the nervous system that controls unconscious or involuntary activities, such as the action of the heart and the digestive system

A2 (a) Noradrenaline
(b) Increased
(c) Increased

A3 (a) Acetylcholine
(b) Decreased
(c) Decreased

***examiner's* note** Some of the functions controlled by the autonomic nervous system can be learned over time and brought under the control of the voluntary nervous system, e.g. bladder control.

A2 **Biology**
The nervous system

The central nervous system

Q1 Name the two components of the central nervous system.

Q2 State one function for each of the following:
(a) hypothalamus
(b) medulla oblongata

Q3 Identify the parts of the brain that carry out the following functions:
(a) balance and posture
(b) voluntary activities, learning and memory

ANSWERS

ANSWERS

A1 The brain and spinal cord

A2 (a) Homeostasis
(b) Controlling heart and ventilation rates

A3 (a) Cerebellum
(b) Cerebrum

***examiner's* note** The spinal cord is the part of the central nervous system that is enclosed by the backbone. It consists of a central cavity containing cerebrospinal fluid, surrounded by a core of grey matter (non-myelinated neurones) and an outer layer of white matter (myelinated neurones). The white matter contains numerous longitudinal neurones which conduct impulses to and from the brain.

A2 Biology
The nervous system

Behaviour: innate

Q1 Distinguish between innate and learned behaviour.

Q2 Distinguish between a taxis and a kinesis.

Q3 What is the term given to sequences of behaviour that control stereotyped responses such as courtship rituals?

ANSWERS

ANSWERS

A1 Innate behaviour is programmed at birth and is genetic in nature. Learned behaviour is acquired during the lifetime of the organism through interaction with the environment.

A2 A taxis is a directional movement in response to a stimulus — for example woodlice show negative phototaxis, i.e. they move away from light. A kinesis is a change in the rate of movement in response to a stimulus — for example sea anemones move their tentacles more when stimulated by chemicals emitted by potential prey.

A3 Fixed action patterns

examiner's note Innate behaviours are usually simple, unchanging responses to certain stimuli. They are programmed at birth and are genetic in nature. Most innate behaviours have obvious benefits in terms of survival or increasing reproductive success.

A2 Biology
The nervous system

98

Behaviour: learned

Q1 Describe what is meant by the term habituation.

Q2 Distinguish between classical and operant conditioning.

Q3 Explain insight learning.

ANSWERS

A1 Learning not to respond to a neutral stimulus

A2 Classical conditioning is learning to associate a neutral stimulus with an important one, e.g. when dogs learn to associate a ringing bell with the arrival of food. Operant conditioning occurs when an animal actively learns to associate an action with a reward or punishment.

A3 Insight learning involves solving a problem by looking at it, thinking about it and using previous experiences to help solve it. It is generally thought that only great apes and humans are capable of insight learning.

***examiner's* note** Although learned behaviour is acquired through experience and interaction with the environment, it does rely on having certain innate abilities. Animals (and humans) will learn some things more easily than others, depending on their innate attributes and cognitive abilities.

A2 Biology
Nutrition

Malnutrition

Q1 Name the two deficiency diseases associated with diets that are low in protein.

Q2 Identify the vitamins associated with these deficiency diseases:
(a) xerophthalmia
(b) scurvy

Q3 Name the functions of the following substances in the human body:
(a) vitamin D
(b) iron

ANSWERS

A1 Marasmus and kwashiorkor

A2 (a) Vitamin A
(b) Vitamin C

A3 (a) To aid the absorption of calcium
(b) To make haemoglobin

***examiner's* note** There are various deficiency diseases associated with malnutrition. In order to remain healthy, people need a balanced diet. In humans, a balanced diet contains proteins, carbohydrates, lipids, vitamins, minerals, water and dietary fibre in the required amounts.

A2 Biology
Nutrition

Overeating

Q1 Name the diseases associated with obesity.

Q2 Insert the missing words: 'Heart disease is often caused by .... Fatty deposits called .... narrow blood vessels and increase .... If this happens in the .... arteries, it can cause angina. Blood clots associated with this problem can cause myocardial infarction (heart attack) and/or .... (bleeding in the brain).'

Q3 What is the main cause of colon disease?

ANSWERS

A1
- Heart disease
- Breast, cervix and colon cancer
- Type 2 diabetes
- Hypertension
- Stroke
- Osteoarthritis

A2 atherosclerosis, atheromas, blood pressure, coronary, strokes

A3 A lack of dietary fibre

***examiner's* note** Obesity is treated by exercising correctly and reducing energy intake. Obese people are encouraged gradually to reduce their intake of energy dense foods, i.e. foods rich in carbohydrate and lipids, while maintaining a reasonable intake of nutrient dense foods, i.e. foods containing protein, vitamins and minerals, in order to maintain a balanced diet.